Eons

raison d'être

By

Damon Dion Reed

Time Created by Thought, i.e. Contents

Chapter 1: Greetings & Salutations

If the universe exists because of energy conservation, then there must be a METHOD to this energy conservation. And if there is a **method** to this energy conservation, then it can be empirically determined. Unfortunately, I'm not in the business of producing empirical data. I'm in the business of producing theories, which helps researchers know where to look for empirical data "patterns". In any event, 'Greetings & Salutations'.

Previously via my imagination, I invented a new language to describe energy conservation, i.e. Quanta Dynamics. Quite simply, Quanta Dynamics is the study of how matter associates based upon the rate and type of degradation. All of which, is based upon 'charge'. Therefore, before I delve into the specific minutia of energy conservation, let me give a quick review.

Thus far, the apex of my 'unified theory' is the Isotopic Degradation Pathway, which connects the behavior of planets/particles and elements/environments. Specifically, different branches of the universe have different charge ratios, which determines the abundance and type of plasma. And as a result of abundance and type plasma, certain electronic structures are supported within atomic particles, which determines the type of atomic interactions, e.g. the exchange of negative Neuprotrons via different electronic layers. All of which, gives a method to the forging of un-imaginable amounts of variation within unique branches of the universe, which have variable amounts of positivity/negativity.

In conclusion, the universe is based upon energy conservation, which means it's highly-illogical that DESTRUCTIVE explosions are conducive to fundamental basis of the universe, i.e. energy conservation. Or in the simplest terms, MOST stars don't go boom; they go snap'crackle'pop until they cool down and produce heavier elements. All of which, leads me to my next great postulate: The heart of the universe is still thumping.

Chapter 2: The Great Thumping

Spontaneous energy appearance has been uncouth for about a century or so...depending on your WiFi availability. Granted, the 'origin' of the initial spontaneous energy appearance is still debatable, i.e. God or happenstance, but we all agree we exist, therefore energy was created. All of which, leads to a pedantic argument of what happened between 'the beginning' and 'us existing'. Personally, I think creation was more methodological than a couple nano-seconds of matter creation, but I'll get to that later. First, let's imagine the heart of the universe is still thumping.

So here is some more review...if you're new to all this hullabaloo. The 'Big Bang' purports a one-time event that will expand until it contracts back to the same point, which I've disagreed with in previous books. In any event, I theorize that God's pet-singularity is still alive.

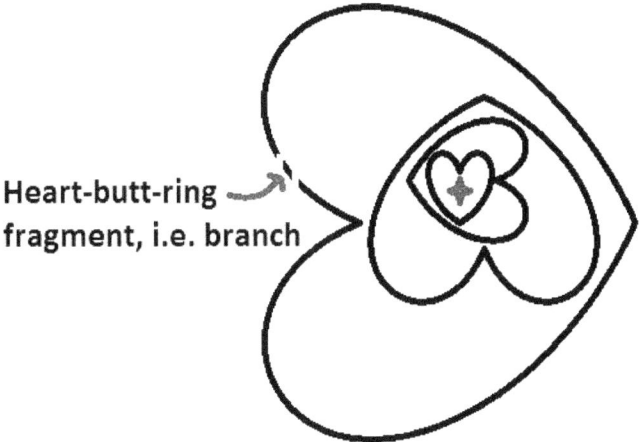

Figure 1: Heart-Butt-Rings

As you can see in this figure, how you look at a problem depends on how you imagine it. And since I've gotten a lot of shit from stupid hearts, I imagine that hearts look like butts. But more importantly, if the blue-X represents God's pet-singularity, then the slow degradation and expansion of heart-butt-rings will make it almost impossible to identify the center of the universe, i.e. the blue-X. Or in other terms, EACH heart-butt-ring contains trillions of unique branches, i.e. unique regions of positivity/negativity. Or in childish terms, as a result of 'wind' blowing differently from each heart-butt-ring, directionality is kind of difficult to establish between Earth and God's pet-singularity.

So where does this leave us...I mean, where does this leave me? Well previously, I used childish words like star-seeds, galactic-seeds, and branch-seeds. (I don't know if I used branch-seeds, but it sounds like something I would say.) Butt now that I'm 40 years old, I'm going to use manly words like: EXO-matter.

After enjoying a moment of profound grunts and groans, being vocalized by the learned members of our community, exo-matter is relatively simple. Matter is CALLED matter because it exists for a VERY-VERY long time. Therefore, exo-matter is stable energy that lasts LONGER than the energy in matter. Or in simpler terms, exoskeletal-matter is the energy in stars, which originated from the exo-matter that forged galaxies, which originated from the exo-matter in heart-butt-rings, which originated from God's pet-singularity. As you can tell, I need a little more nomenclature than just exo-matter. Therefore, exo(s)-matter forges stars, exo(g)-matter forges galaxies, and exo(hbr)-matter forges heart-butt-ring branch fragments, i.e. Wiley Coyote's TNT fuse, and exo(PS)-matter represents God's pet-singularity. Soooo, what is Wiley Coyote's TNT fuse?

God's pet-singularity is degrading these heart-butt-rings of extremely complex, highly energetic, and uniquely charged exo(hbr)-matter fragments. All of which, have the propensity to uniquely associate to **delay** their <u>degradation</u> into unique branches, i.e. Wiley Coyote's TNT fuse. Or

6

in simpler terms, exo(hbr)-matter can form complex 'element' arrangements…to prolong their decay. And if that doesn't make your head hurt, just imagine these complex 'element' arrangements of exo(hbr)-matter might be exchanging exo(g)-matter via Neuproz-esk behavior, **i.e. really confusing!** (FYI, science-NERD is my primary language.)

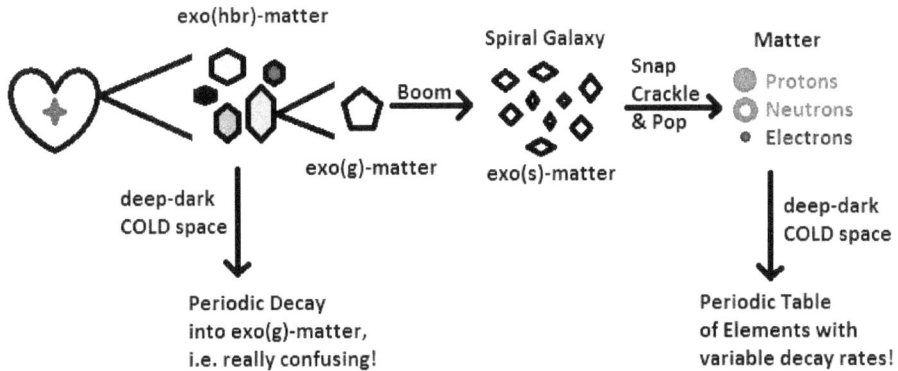

Figure 2: One CONFUSING Thump

Now what you can't see in this figure is: how the unique association of exo(hbr)-matter **delays** the degradation into exo(g)-matter. Butt, it is definitely implied based upon the statement of "i.e. really confusing". Or in simpler terms, the **DELAY** is based upon the exchange of exo(g)-matter between chunks of exo(hbr)-matter.

So let's recap. Based upon the energetic complexity of protons, neutrons, and electrons, it is logical that matter stemmed from exo-matter, which has more complexity and lasts LONGER than matter. Also, the unique association of exo(hbr)-matter can result in branches with unique rates of decay. All of which, is complicated by the expansion of energy based upon the energy-environment and/or the age of the energy. Or in simpler terms, it is possible that a piece of exo(g)-matter, which will produce a spiral galaxy, is smaller than a proton. (Brain Ka-boom…ahhh, thank you very much.)

In conclusion, the universe is based upon energy conservation and big booms are not conducive to energy conservation. Therefore, by working backwards, it seems logical that God's pet-singularity decayed really-REALLY slowly. Butt, each time God's pet-singularity degrades a little, it creates a heart-butt-ring, which contains exo(hbr)-matter. As a result of the association of exo(hbr)-matter, the rate decay is based upon the environment and electronic order. All of which means, exo-matter lasts longer than matter, which means God's pet-singularity is probably thumping out heart-butt-rings...thankfully?

Chapter 3: Timeworms

Honestly, I believe in wormholes. But, I believe in **wormholes** ONLY because I believe in <u>Timeworms</u>. For those of you who are unaware of Timeworms, they look like that weird-dragon-thingie from *The Never Ending Story* that wiggles through the air, which is kind of ironic since Timeworms are not constrained by space and time...let alone air. In any event, I believe that Timeworms have existed since before time, throughout time, and are currently fly through space-time...munching the energy between different branches of the universe to create wormholes, which do NOT adhere to the entropic rules of the universe.

Now don't get me wrong, I truly enjoy the presence of science in science-fiction because it does a great job of spurring the common folk into enjoy consciousness for the briefest of moments. Butt, things have gotten out of hand. Therefore, let's take a moment and add a bit of realism to our imagination. (The reason why things have gotten out of hand is because we don't have enough scientists in Congress?)

For those of you that are furious with regards to my sarcasm, let me be completely clear: Rarefactions about the plasma-energy that fills the space between planets, stars, and galaxies...is possible. But, in as much as it is a tubular VOID of energy that can appear and suck things from one side of the universe to another, without applying any physics of time or space, is a bit of a stretch. Or in Quanta Dynamic terms, rarefactions about the energy being released by exo-matter does NOT mean a firehose will SPONTANIOUSLY appear about the rarefactions.

In conclusion, **unless** there are Timeworms eating at the core of a Time-apple, then wormholes are simply an 'addictive idea' that has captivated humanity for too long. (Unfortunately, addictive ideas are a common occurrence to humanity.) Fortunately, since most people get their apples from supermarkets these days, there is less of a chance of finding a worm in your apple, which will allow the next generation of scientists to release the idea of 'wormholes' to **diffuse** into the *Never Ending Story*, e.g. universe as we know it.

Chapter 4: Just Sparks

Now that we all agree that 'future' scientists will not partake in the wormhole addiction, the only thing that remains is to ponder the interaction between different branches of the universe, i.e. unique regions of God's thumping heart. Or in simpler terms, we might only be able to survive in the Milky Way.

In *Einstein Asymptote*, I made the postulate that the most abundant plasma fragment, e.g. thermal energetic quanta, intercalates and determines the volume of matter. Now, just to clarify, there are MORE abundant and MUCH smaller plasma fragments than thermal energetic quanta. But these SMALLER energy fragments don't respond to most of gravity's factors, which I'll explain more later. In any event, with that clarification out of the way, let us return the uniqueness of a spiral galaxy.

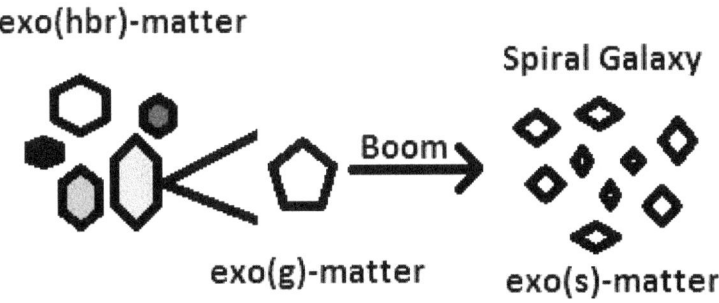

Figure 3: Exo(g)-matter

As you can see in this figure, the unique decay of the exo(hbr)-matter results in unique exo(g)-matter, which goes boom into a spiral galaxy. Therefore, even though spiral galaxies within a given branch of the universe have similar overlaps to allow for the transmission of light between spiral galaxies, it is UNCLEAR if each spiral galaxy has the same thermal energetic quanta. And since the thermal energetic quanta facilitate volume in this negative branch AND volume plays an integral part in the formation and behavior of elements, a slight variation in volume will have a drastic effect. Granted, those drastic effects are simply bond-angles, but we're rigidly BUILT upon **specific** bond angles. So you know, things would really get fucked up if our DNA were distorted...just a little bit. All of which, brings us to the title of this chapter: Just Sparks.

For the most part, based upon the vastness of time across all levels of matter and exo-matter as well as the constant diffusion of our heart-butt-ring, we are bound to observe some light NOT created by stars. But before we return to the wormhole-addiction, let's remember that we can't see everything...especially if the interaction is between different branches of the universe.

In conclusion, there is a remote possibility that God didn't create all thermal energetic quanta equally. Or in scientific terms, the Milky Way might have unique thermal energetic quanta, which determines the unique ANGLES in our DNA. Or in simpler terms, the Milky Way may be the only place where we can exist...God, I wish it was bigger!

Chapter 5: Endo-matter

Over the last few chapters, I've theorized such great things as: God's heart still thumping, exo-matter, and our galactic isolation, which isn't really a problem right now. All of which, leads to the following question: How can you tell the difference between exo-matter and endo-matter? And more importantly, what is endo-matter?

Previously, I postulated second-order-matter can be forged in black holes, which will either explode or give rise to a second-order-black hole OR galaxy. Or in scientific terms, the fusion of extremely negative energy within the Milky Way's black hole will give rise to endo-matter. Or in simpler terms, endo-matter is stable energy created by the degradation of matter in a given branch of the universe to give a uniquely different negative/positive electronic entity. (It is almost as if these abbreviations were invented just for Quanta Dynamics;-)

All of which, brings us back to the following question: How can you tell the difference between exo-matter and endo-matter? Quite simply, it's really tough. Both will appear to be black holes, i.e. 'matter' destroying entities. Now, based upon the fact that endo-matter is LESS energetic, it would seem logical that endo-matter black holes will be smaller and have shorter lives. But, being that endo-matter is a derivation of matter, which expands based upon the presence of thermal energetic quanta within a given galaxy, exo-matter might be smaller. Thankfully, the only way to tell the difference, without empirical data, is by the amount of destruction they do when they go BOOM. In any event, endo-matter sounds like a much more mature term.

With all that nomenclature out of the way, it is possible to create endo-matter on Earth? And if so, would it have some beneficial attributes? And finally, who is going to freak-out over this concept?

Well, to make endo-matter on Earth, you'll need to squish electrons together until they degrade and start sharing some negative energetic quanta. All of which, can only be accomplished with super-strong negative magnetic environments. Now, what are the benefits of creating endo-matter? The unique degradation of this endo-matter will probably result in destruction of background plasma, which will produce some thermodynamically interesting events. Or in simpler terms, we get most of our energy from 'thermodynamically interesting events'. Unfortunately, everyone is afraid of black holes. (It's a black thing...I guess?)

Everybody knows that black holes are Godless killing machines that do NOT adhere to normal physics. All of which means, if a black hole is created, NO MATTER THE SIZE, pun intended, it will become the BLOB! Yes, I said it, the BLOB. And as everybody knows, the BLOB has absolutely NO limitations to its energy and will grow forever and ever and ever...Amen. Or in less sarcastic terms, black holes do **NOT** contain an **infinite** amount of energy that will grow uncontrollably until all of humanity is dead, i.e. the military industrial complex. Or in less whiny political genocidal terms, creating an endo-matter black hole with several electrons WILL only have the energy of SEVERAL electrons. All of which means, the chance of catalytic cataclysmic event resulting in all matter on Earth degrading...is insanely slim to nil. Or in the simplest terms, MATTER HAS BEEN GOING BOOM SINCE THE BEGINNING OF TIME AND that electron that you feasted on last week, has probably already died a horrible-horrible death. And guess what? The universe didn't end. Imagine that?

In conclusion, Quanta Dynamics not only does a tremendous job at methodologically explaining the universe, which is based upon energy degradation as a function of conservation, it will also allow scientists to experiment on creating futuristic energy sources without shitting their pants. Granted, common folk will probably still shit their pants, but pooping once a day is my mantra. And finally, I have no clue on how to figure out the difference between exo-matter and endo-matter, but creating endo-matter would be very interesting. (Truthfully, I just want everybody to feel my intestinal pain...insert evil laugh here:-)

Chapter 6: Not Again...(sigh)

Welcome back to my reoccurring literary segment called: How much of the universe can we really see? (I tried to think of a better name, but I totally drew a blank.) Previously, I postulated that light energy is only stable within a given branch of the universe, which has a unique amount of positivity/negativity. But, with the theorization of exo & endo matter as well as heart-butt-rings irradiating from God's pet-singularity, I think it is about time to take another look:-)

exo(hbr)-matter

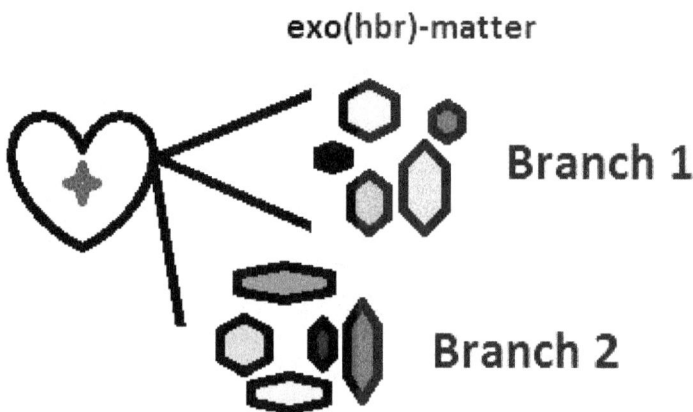

Figure 4: Two Branches

As you can surmise from this picture, hopefully, is that unique Branch 1 is NOT visible to unique Branch 2. But, how does our universe look as a result of Branch 2 going BOOM and fucking up the dispersion of Branch 1? Well, let's imagine.

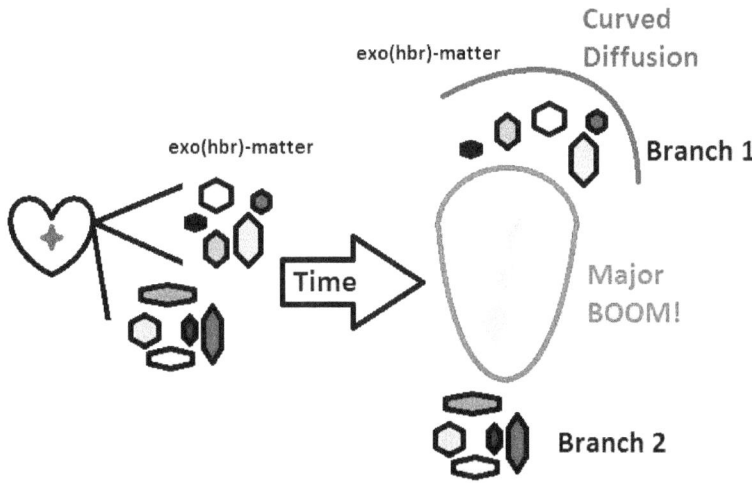

Figure 5: Curved Diffusion

As you can see in this figure, a hypothetical 'Major BOOM' has caused Branch 1 to adopt a curved nature. Now in all honesty, this example is a tad bit simplistic in that the heart-butt-ring probably has multiple branches, i.e. clusters of exo(hbr)-matter. Also, there is a chance of a completely UNIQUE CROSS-heart-butt-ring occurrence.

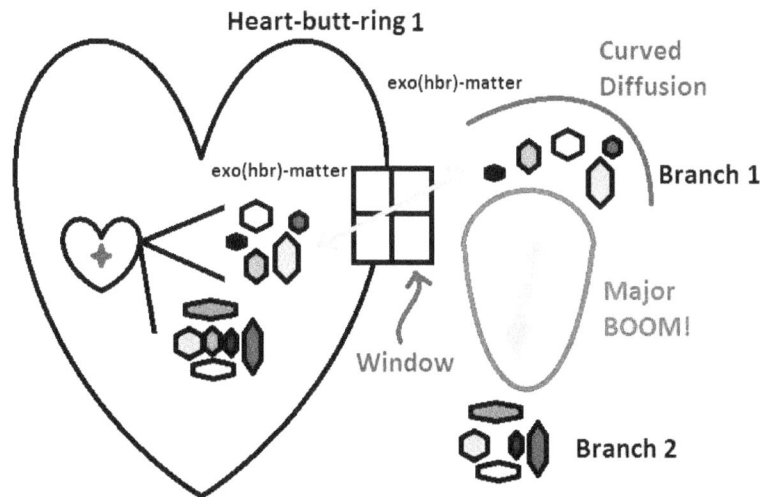

Figure 6: A Spectacle to SEE

As you can see by the double ended arrow shooting across a window between heart-butt-ring-1 and the smaller heart-butt-ring-2, there is a FREAK-similarity between the photons of these two branches, which allows people in heart-butt-ring-1 to see people in heart-butt-ring-2. All of which, makes me wonder the following: Is the diffusion of heart-butt-ring-1 different than heart-butt-ring-2? The short answer is: Yes. The long answer is: Wind shield wiper example in my book: *PLINTH*, B.U.T.T., chapter 3.

In conclusion, I ALWAYS assume I don't have a firm grasp on the complexly of the universe. And as a result of this, I try never to make gross over-simplifications based on what is visible to our wimpy telescopes. And finally, it's freaky that we exist. Therefore, I assume that other freaky things exist as well.

Chapter 7: Protective Crevices

I know it's difficult to imagine, but science has a flaw: Environments. Or in simpler terms, there are HUGELY specific factors that need to be held constant about elements to detect specific energetic quanta. Granted, I've already mentioned the fact that super colliding magnets create negative environments, which unduly INFLUENCES the detection of certain energetic quanta, but now I'm going to take it a step further. Actually, I already took it a step further in *Asymptote Einstein*, so I'll have to take it even further. But first, a quick review.

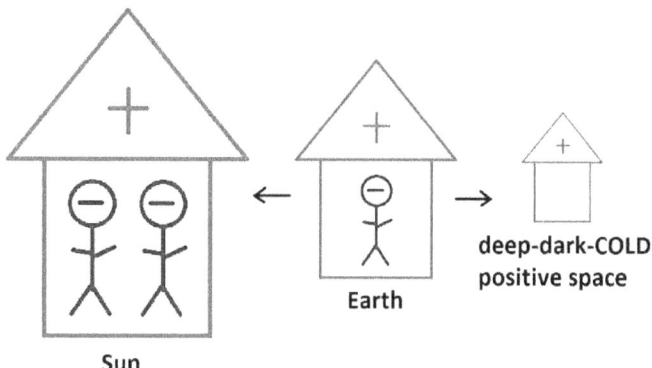

Figure 6: Our Branch

In our branch of the universe, environment is everything: **Excess** negative-heads causes positive-matter-houses to **expand**; absence of negative-heads causes positive-matter-houses to shrink. It is pretty simple if you think about it. But, what isn't simple is that there is a spectrum of positive-matter-houses and negative-heads, which creates a spectrum of relative half-lives. Or in simpler terms, look down.

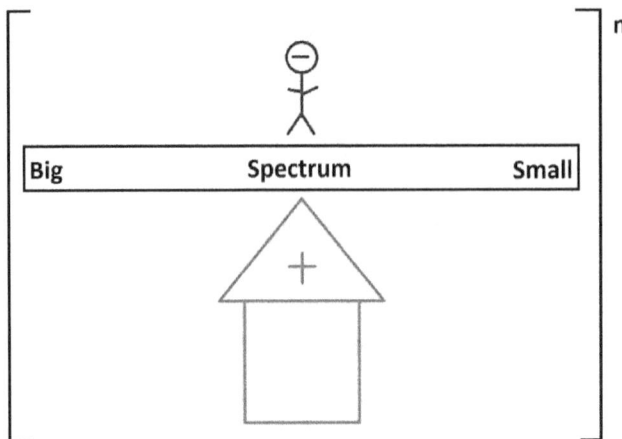

Figure 7: Spectrum

EVERY unique energetic quantum can exist in various states: Big or Small. But, every unique energetic quantum has a different half-life based upon its location about the big/small spectrum. And to make things even more difficult, different stars release different spectrums of negative plasma, which modulates each spectrum.

With all this in mind, it is NOT outside the realm of possibilities that different stars impregnate 'matter' differently, which will modulate the half-life, but also the gravitational response. For example, look down.

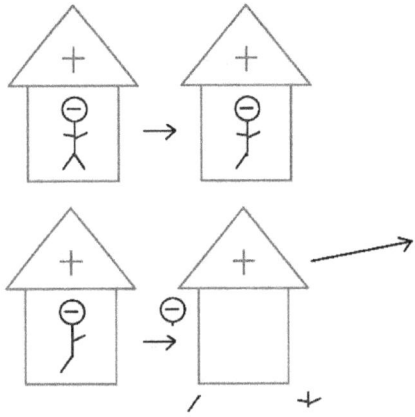

Figure 8: Negative-Head Pop

As you can see, due the fact the negative-head with one leg/arm went 'pop' sooner, which that proton displayed more positive charge. And as a result of having more positive charge, it is attracted MORE to the negativity of the sun, i.e. gravity.

For a couple hours, I tried to imagine a good way to explain ALL the factors that play into the half-life of energetic quanta, matter, & PLASMA. And here is what I got.

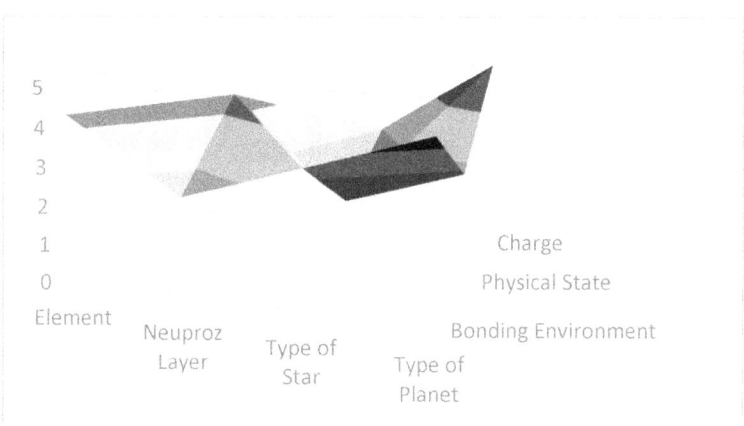

Figure 9: Complex

Hopefully, you can surmise that this figure does NOT correlate to a unique energetic quantum, but it simply represents the congealing of ALL the factors that play into half-lives. For example, the overall atomic orbital arrangement, i.e. element, is a factor; Neuproz Layer determines the depth of the crevices; The type of star determines the underlying abundance of certain plasma; The magnetic environment determines atomic orbitals and abundance of certain plasma; Bonding environment plays a part in vibrational distortion of atomic orbitals; Physical state affects the directional exposure of atomic orbitals, i.e. rotation; And finally, the charge determines the amount of attraction between energy. Like I said, almost every single factor exists as a spectrum and plays a part in determining the half-life of an ENERGETIC quantum. Or in scientifically disturbing terms, it is possible that there are bits of plasma wedged within a proton on Earth that last LONGER than an **actual** proton lasts in deep-dark-COLD positive space. (Thank you Mr. Sun for your patient environment?)

In conclusion, I totally support the work of scientists within super-scientific super-colliders, but I would advise that caution be placed on the wording of your educational endeavors, i.e. please make it apparent that half-lives of energetic quanta are RELATIVE to the environment. Also, it would be nice if you OPENLY admitted where progress can be made, because it gives youthful researchers hope with regards to notoriety, which is a HUGE factor in the amount of HEART researchers put into their work. And finally, as a result of the universe's complexity, the minutia surrounding the smallest of things can be tremendously relevant.

Chapter 8: Electro-tivity

In as much as 'Noble gasses' prefer not to get involved in the sharing of electrons, the rest of the periodic table displays a spectrum of behaviors. The left side contains 'electropositive' elements and the right side, near the Noble gasses, contains 'electronegative' elements. All of which, is the result of some very interesting weirdness. Therefore, to begin this journey of weirdness, let's review the periodic table and the isotopes therein.

^{1}H 1,2,3							Atomic Number → ^{15}P Isotopes										^{2}He 3,4
^{3}Li 6,7	^{4}Be 9											^{5}B 10,11	^{6}C 12,13, 14	^{7}N 14,15	^{8}O 16,17, 18	^{9}F 19	^{10}Ne 20,21 22
^{11}Na 23	^{12}Mg 24,25, 26											^{13}Al 27	^{14}Si 28,29, 30	^{15}P 31	^{16}S 32,33, 34,36	^{17}Cl 35,37	^{18}Ar 36,38, 40
^{19}K 39,40, 41	^{20}Ca 40,42, 43,44, 46,48	^{21}Sc 45	^{22}Ti 46,47, 48,49, 50	^{23}V 50,51	^{24}Cr 50,52, 53,54	^{25}Mn 55	^{26}Fe 54,56, 57,58	^{27}Co 59	^{28}Ni 58,60, 61,62, 64	^{29}Cu 63,65	^{30}Zn 64,66, 67,68, 70	^{31}Ga 69,71	^{32}Ge 70,72, 73,74, 76	^{33}As 75	^{34}Se 74,76, 77,78, 80,82	^{35}Br 79,81	^{36}Kr 78,80, 82,83, 84,86
^{37}Rb 85,87	^{38}Sr 84,86, 87,88	^{39}Y 89	^{40}Zr 90,91, 92,94, 96	^{41}Nb 93	^{42}Mo 92,94, 95,96, 97,98, 100	^{43}Tc 98	^{44}Ru 96,98, 99,100, 101,102, 104	^{45}Rh 103	^{46}Pd 102,104, 105,106, 108,110	^{47}Ag 107,109	^{48}Cd 106,108, 110,111, 112,113, 114,116	^{49}In 113,115	^{50}Sn 112,114, 115,116, 117,118, 119,120, 122,124	^{51}Sb 121,123	^{52}Te 120,122, 123,124, 125,126, 128,130	^{53}I 127	^{54}Xe 124,126, 128,129, 130,131, 132,134, 136
^{55}Cs 133	^{56}Ba 130,132, 134,135, 136,137, 138	Lantha- nides	^{72}Hf 174,176, 177,178, 179,180,	^{73}Ta 180,181	^{74}W 180,182, 183,184, 186	^{75}Re 185,187	^{76}Os 184,186, 187,188, 189,190, 192	^{77}Ir 191,193	^{78}Pt 190,192, 194,195, 196,198	^{79}Au 197	^{80}Hg 196,198, 199,200, 201,202, 204	^{81}Tl 203,205	^{82}Pb 204,206, 207,208,	^{83}Bi 209	^{84}Po 209	^{85}At 210	^{86}Rn 222
^{58}Ce 136,138, 140,142	^{59}Pr 141	^{60}Nd 142,143, 144,145, 146,148, 150	^{61}Pm 145	^{62}Sm 144,147, 148,149, 150,152, 154	^{63}Eu 151,153	^{64}Gd 152,154, 155,156, 157,158, 160	^{65}Tb 159	^{66}Dy 156,158, 160,161, 162,163, 164	^{67}Ho 165	^{68}Er 162,164, 166,167 168,170	^{69}Tm 169	^{70}Yb 168,170, 171,172, 173,174, 176	^{71}Lu 175,176				

Figure 10: Spectrum Table, a.k.a. Periodic Table. (From *Incorporeal.*)

As you can see in this figure, weirdness is pervasive. Or more specifically, I have it on good authority that the SPECTRUM of 'electro-tivity' that spans across the periodic table, electropositive on the left and electronegative on the right, ALSO spans the isotopes. Or in simpler terms, Zinc's Isotopic electropositivity is as follows: 64>66>67>68>70. Or in other terms, Zinc's Isotopic electronegativity is as follows: 70>68>67>66>64. I guess any way you look at it, there is a SPECTRUM. But, what are the factors of this spectrum? Is the spectrum based solely on the Neuproz Groupings and the distortion caused by the placement of the extra neutrons? Or, is the spectrum simply an atomic orbital thang? (BTW, both answers are interrelated and I'm surprised the word 'thang' is in the Word Dictionary.)

For a second, let's **forget** everything we know about science. (Easier for some people in comparison to others.) Now, let's imagine there's a spectrum of atomic nuclei radii within the periodic table: The atoms on the left have big radii and the atoms on the right have smaller radii. (This is actually a scientific fact, but don't tell anyone.) With that said, electro-tivity is FORMALLY an electron thang. (Damn, the word 'thang' is fun to use.) But, when you look closer at an element's ability to hold onto electrons, it becomes apparent that electo-tivity is a function of the atomic nucleus, atomic orbital connectivity, and the environment. All of which, gets really weird, REALLY QUICK, in the absence of Quanta Dynamics. Thankfully, I invented Quanta Dynamics, so let's use that to avoid some of the weirdness.

In a Quanta Dynamic universe, electro-tivity is determined by the amount of negative plasma surging around an atom. Or in simpler terms, electronegative elements SUCK-UP more negative plasma than electropositive elements. All of which, is based upon the atomic orbitals and everything that correlates to atomic orbital placement. Or in the simplest terms, electronegativity is caused by the Admiral Ackbar Theory. Therefore, to understand electro-tivity, all we have to do is think about the factors associated with the Admiral Ackbar Theory.

The premise of the Admiral Ackbar Theory is: the SPACE between atomic orbitals allows negative plasma to **surge** towards the POSITIVE atomic nucleus before being swatted away by the movement of electrons. Or in figurative terms, look down.

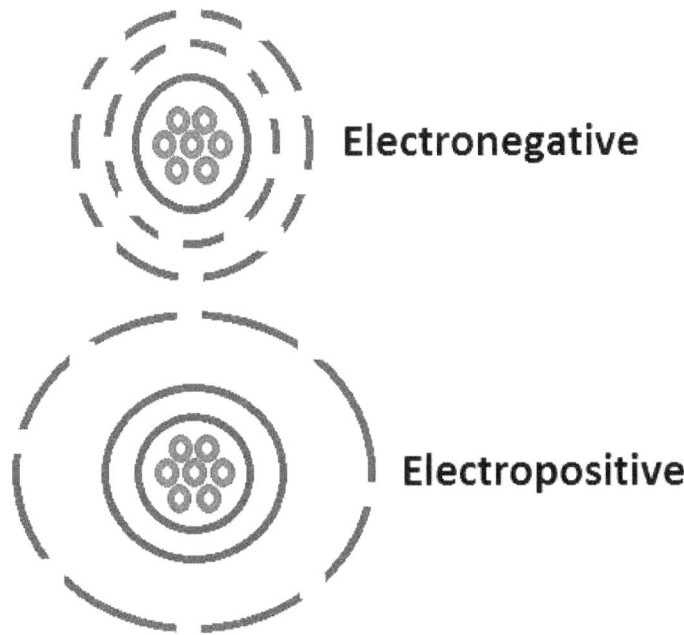

Figure 11: Electropositive Vs Electronegative

As you can see in this figure, based upon the "holes" in the atomic orbital shells, negative plasma can surge closer to the positive atomic nucleus of the of electronegative atoms. Conversely, based upon the "lack of holes" in the atomic orbital shells, electropositive atoms hold negative plasma further away from the positive atomic nucleus. All of which, is formally confusing when you remember the 'Protective Crevices' chapter. Therefore, let me try and explain.

The atomic orbitals of electropositive atoms enlarge as a result of negative plasma, i.e. thermal expansion. As a result of this, certain elements in certain environments will release their valence electrons. But, as a result of this shift in atomic orbital placement, it also '**closes**' the inner atomic orbitals, which inhibits the influx of negative plasma towards the atomic nucleus.

On the other hand, electronegative atoms in certain environments will allow more negative plasma closer to the positive atomic nucleus, which means the constant movement of negative plasma around electronegative atoms will increase the negativity of the atom, i.e. electronegative. Or in other words, the increase of extra neutrons results in LESS order with regards to atomic orbital placement, which allows for more room for surging negative plasma towards the positive atomic nucleus, i.e. electronegativity. (Totally weird that extra-neutrons are the determining factor in electronegativity?) In any event, let me take a moment and add on one more level of complexity.

Quite simply, the placement of the extra neutrons about larger Neuproz Groupings determines the electro-tivity because larger Neuproz Groupings are more energetic. Or in other terms, larger Neuproz Groupings are exchanging negative Neuprotrons via deeper layers, which means the negative Neuprotrons are moving faster. And as a result of the negative Neuprotrons moving faster, their averaged 'presence' will have more of an influence on electron movement and atomic orbital placement. All of which means, the distortion of larger Neuproz Groupings by extra neutrons results in MORE atomic orbital distortion, i.e. electronegativity.

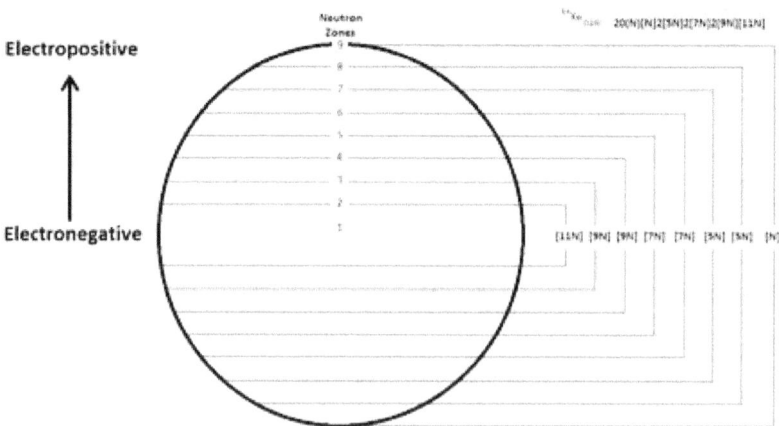

Figure 12: Electro-tivity as a function of Neutron Zones. (From *Incorporeal*.)

As you can see in this figure, the incorporation of an extra neutron in Zone 1 will make the atom more electronegative than an extra neutron in Zone 8, which will make it more electropositive. With all that in mind, I think I need to clarify the METHOD by which the rate of negative Neuprotron exchange between larger Neuproz Groupings results in the placement of atomic orbitals.

Electrons surge towards the atomic nucleus because of the positivity therein. But, inside the atomic nucleus, there are PILLARS of negativity, created by the movement of negative Neuprotrons, which **steer** electron movement. The faster negative Neuprotrons are exchanged between Neuproz Groupings, the larger the negative PILLARS and the more electrons will be steered by the negative PILLARS.

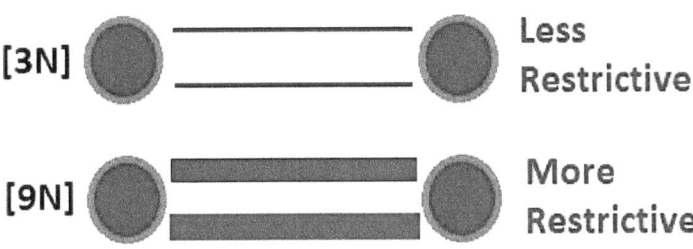

Figure 13: Partial negative charge as a function of exchange rate.

As you can surmise from this figure, the partial negative charge 'pillars' are larger between [9N] groupings because of the **rate** of negative Neuprotron exchange. And as a result of these restrictive negative 'pillars', larger Neuproz Groupings have MORE influence upon electron movement, which determines atomic orbital placement. Or in cross-section terms, look down.

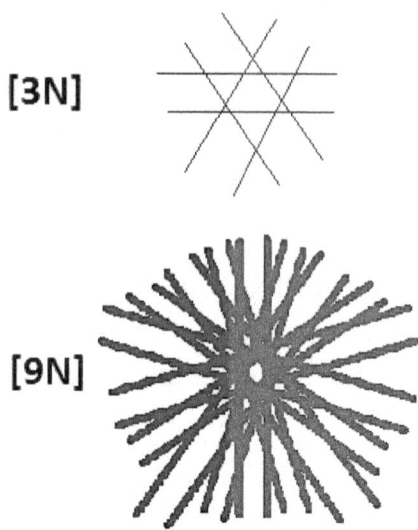

[3N]

[9N]

Figure 14: Steering Holes

As you can see in this figure, the cross-section of a [9N] grouping is more complex and restrictive than the cross-section of a [3N]. Now what you can't see, but you might be able to surmise, is that based up the holes within a [9N] grouping, variance in the external environment has the propensity to push an electron's trajectory such that it flies through a **different** hole. And when an electron goes through a 'different' hole, it creates a different atomic orbital because it experiences a different positive pattern from the atomic nucleus. All of which, is complicated because negative electrons flying through this complex pillar negativity causes vibrations about the negative pillars, which affects the movement of the other electrons. Actually, it's kind of beautiful how columbic dissonance caused by one negative electron can affect the movement of another negative electron. (FYI, if you imagine the overlap of all these 'patterns' created by Neuproz Groupings within the atomic nucleus, you should be able to see the 3-dimensional path by which certain electrons move...to create specific atomic orbitals, but that is for another book. The short and skinny of it all is: Smaller holes LOCK atomic orbitals together, which creates cohesive atomic orbitals that block the surge of negative plasma towards the atomic nucleus, i.e. electro-positivity; Bigger holes allow for atomic orbital wobbling, which allows negative plasma to surge closer to the atomic nucleus, i.e. electro-negativity.)

Another reason why the incorporation of extra neutrons about larger Neuproz Groupings has more of a influence upon electron movement is: more partial POSITIVE charges are smooshed about the axis of the atomic nuclei, which plays a part in the distortion of the negative electrons that create the atomic orbitals.

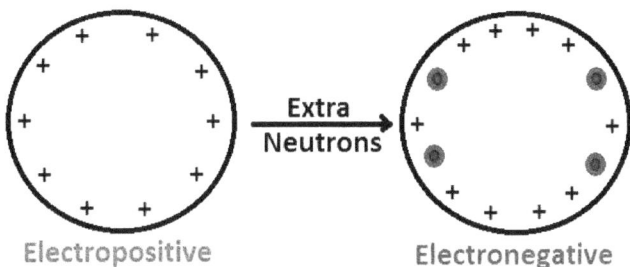

Figure 15: Charge Migration

As you can see, the extra neutrons, which are brown and indigo, focus the positive charges towards the axis of the atomic nucleus. And as a result of this increase of partial positive charge about the axis of the atomic nuclei, negative atomic orbitals migrate toward the axis of the atomic nucleus.

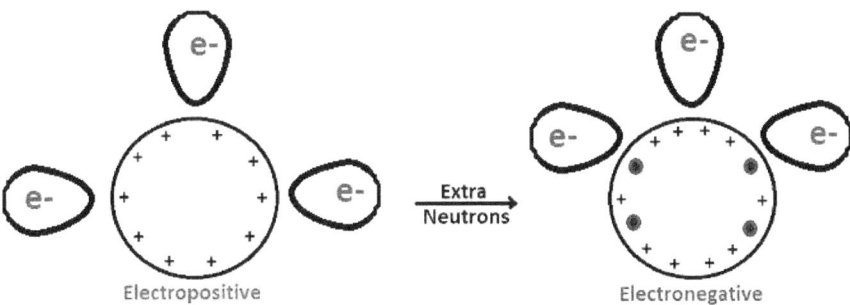

Figure 16: Atomic Orbital Placement

In as much as this figure is an exaggeration, it makes it really easy to see HOW the negative electrons adjust based upon the squishing of partial positive charges towards the axis, which is the METHOD by which more space is imparted **between** atomic orbitals such that negative plasma can flow closer to the positive atomic nucleus, i.e. electronegativity. Unfortunately, there is another weird factor associated with electro-tivity: Twisting.

As a result of there being a spectrum of electro-tivity about the periodic table and isotopes, electro-tivity is a function of neutron placement, which distorts the alignment of Neuproz Groupings. More specifically, as a result of variable neutron placement within stereo-isotopes, it is more than likely that Neuproz Pairs can twist.

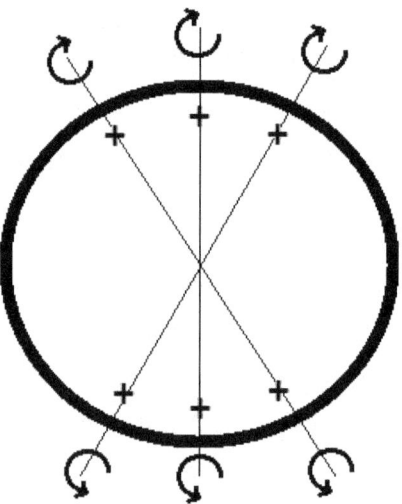

Figure 17: Neuproz Pair Twisting

As you can see in this figure, it is theoretically possible that Neuproz Pairs within a Neuproz Group can twist. All of which means, there is a spectrum of Neuproz Groupings, i.e. $[3N]p_1$, $[3N]p_2$, $[3N]p_3$, which are all based upon the cohesion between different rotational Neuproz Pairs. But, since the cohesion about the atomic nucleus is a function of electronic cohesion, twisting will result in a decrease in cohesion and an expansion of the atomic nucleus. Or in simpler terms, a Neuproz Grouping without any twisting, [3N], has more cohesion than a Neuproz Grouping with one twist,

[3N]p_1, two twists, [3N]p_2, and three twists [3N]p_3…etc. Or in the simplest terms, there are 360 degrees of possible twists for each pair, i.e. a spectrum of cohesions within a single Neuproz Grouping. (FYI, the increase in twists per Neuproz Grouping results in expansion, which facilitates Isotopic Degradation.)

In conclusion, the Admiral Ackbar Theory determines electro-tivity and is a function of extra neutrons. As per the METHOD of electronegativity, it is based upon the postulate that larger Neuproz Groupings are MORE energetic and their distortion will result in greater atomic orbital distortion. Also, extra neutrons about the larger Neuproz Groupings results in the smooshing of partial positive charges towards the axis of the atomic nucleus, which also distorts atomic orbital placement. And as a result of LESS cohesive atomic orbitals, there is MORE room for negative plasma to surge towards the atomic nucleus, which increases the electronegativity. And finally, the twisting of Neuproz Pairs within a Neuproz Grouping is a stepwise method to the decay of Neuproz Groupings, which facilitates the Isotopic Degradation Pathway. (BTW, have you ever noticed that electropositive atoms don't make covalent bonds, which are a function of atomic orbital overlap? Or in simpler terms, there is a spectrum of factors associated with atomic orbital overlap, i.e. covalent bonds, which I'll talk about in another book because it's weirdly complex.)

Chapter 9: Extra Neutrons

As a result of extra neutrons causing the Admiral Ackbar Theory, which increases the rate of decay via the isotopic degradation pathway, it is important to note that 'fusion' has a space-component and older planets have massive amounts of helium in their atmospheres. All of which, results in the postulate that fusion occurs about extra neutrons, which are forced to the axis of the Neuproz Cluster. (From *Incorporeal.*) But, what happens when NON-axial neutron(s) degrades or undergoes fusion to create a NON-spherically orientated Neuproz Pair?

For starters, the degradation of NON-axis extra-neutrons without expulsion from the Neuproz Cluster will formally convert the element up one position in the periodic table, which is weird. For starters, isotopic degradation pathway usually results in the formation of lighter elements via the loss of two negative Neuprotrons and the inclusion of four extra-neutrons. Or in simpler terms, look down.

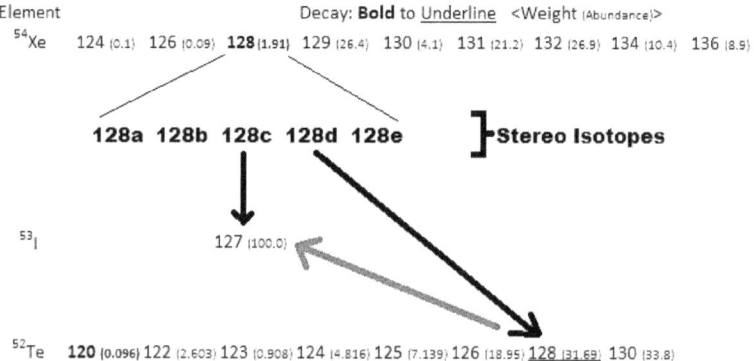

Figure 18: Typical and Odd Degradations

As you can see in this figure, **typical** degradations are in black and <u>odd</u> degradations are in red. Or more specifically, a **typical** degradation is when a stereo isotope of Xenon (128d) degrades to form Tellurium(128) via the decay of two protons into two neutrons. And an <u>odd</u> degradation is when Tellurium(128) loses one neutron and one neutron decays into one proton to yield Iodine(127). All of which, is highly dependent on the environment and stereo isotope. But, the weirdness doesn't stop there.

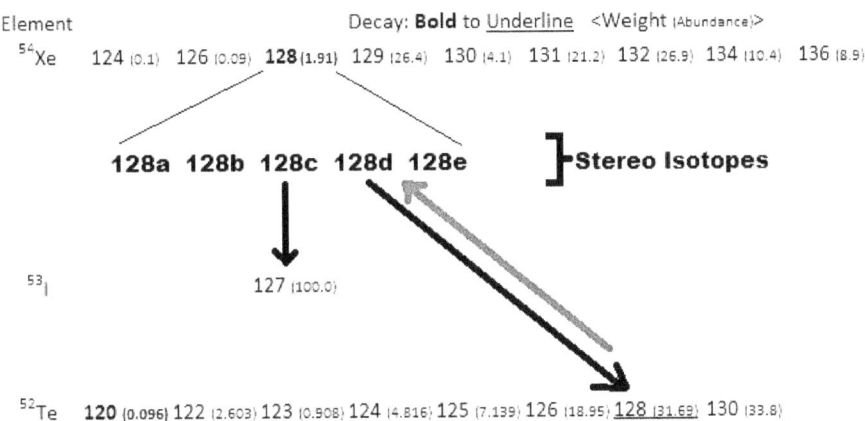

Figure 19: Odder Degradations

As you can see in this figure, it is theoretically possible that the newly excluded four neutrons of Te(128) degrade again to form two NON-spherically orientated Neuproz Pairs, which would formally convert Te(128) back into Xe(128d). But, there will be a tremendous amount of electronic repulsion about these two NON-spherically orientated Neuproz Pairs. Or in simpler terms, on Earth, this ODD degradation is very uncommon because of the electronic repulsion, which is based upon environment. But in deep-dark-COLD positive space, this NON-spherical formation of TWO Neuproz Pairs might be conducive to energy conservation. All of which, makes deep-dark-COLD positive space even weirder. Let me try and explain.

Even though the periodic table exists as a spectrum of stereo isotopes about each isotope, in deep-dark-COLD positive space, there are the possibility of pseudo-elements, which have the requisite atomic numbers, but lack the spherically orientated Neuproz Cluster. Or in simpler terms, stereo isotopes will have similar characteristics based upon slightly different atomic orbital arrangements, but pseudo-isotopes will have surprisingly different characteristics, which will probably baffle scientists and make the periodic table look less periodical.

In conclusion, fusion is a function of proximity and electronic alignment. And based upon the isotopic degradation pathway pushing extra neutrons towards the axis of the Neuproz Cluster, this is a very VIABLE pathway to the extrusion of helium, which is a huge part of older planets. And based upon this logic, and the logic of the previous chapter, electro-tivity is a function of neutron placement, but the behavior of extra neutrons is dependent on the environment. Or in simpler terms, variable negativity will determine the volume of the extra-neutrons within the Neuproz Cluster, which determines if extra-neutrons form oddly NON-spherical Neuproz Pairs. All of which, will modulate electron placement and elemental characteristics, i.e. pseudo-isotopes with really weird characteristics.

Chapter 10: Understanding Fission

Another function of proximity is: Decay. Unfortunately, since we exist in a universe of moving parts, vibration is also a function of decay. All of which, is a function of electronic meshing between Neuproz Particles, which is based upon the external environment. Therefore, let's take a moment and think about all this complexity as it relates to understanding fission...if that sort of thing interests you.

For those of you who don't know this, atomic energy is derived via one specific isotope of Uranium, which is really interesting. Therefore, let's think about the **OLD** theory of fission.

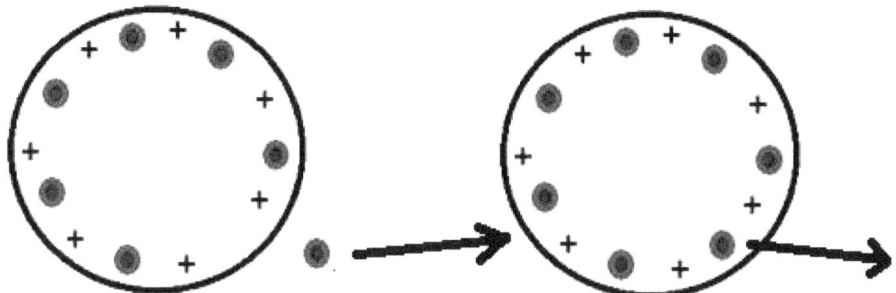

Figure 20: Uranium Fission

As depicted by this figure, the old theory of fission purports that a released neutron crashes into another atomic nucleus, which causes the second atomic nuclei to release another neutron. Pretty simple, right? Well, not so much. There is one major problem with this OLD theory: negative thermal energetic quanta.

To any fission-ary novice, if an atomic reactor gets too hot, there will be a meltdown...right? (Thank you Simpsons for helping society know one thing with regards to fission-ary science.) In any event, this obscure fact results in the downfall of the old theory of fission. Or in simpler terms, take a gander at the next figure for a moment.

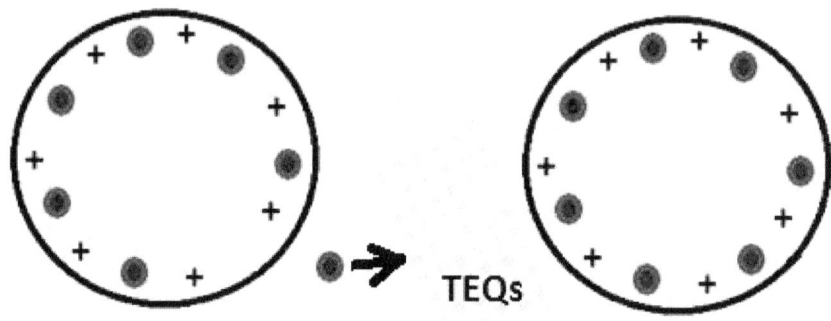

Figure 21: TEQ Problems

In theory, the MASSIVE amount of TEQs should prevent the released-neutron from bumping into the adjacent uranium atomic nucleus...to cause the cascading reaction. And just in case you weren't convinced that about the OLD-theory being substandard, reactor meltdowns create different radioactive waste. Or in simpler terms, thermal energetic quanta are an INTEGRAL part in the cascading reaction, which means the OLD theory is shit. I mean, how is NEUTRON collision the ONLY factor in uranium decay? In any event, here is my **new** theory.

Uranium releases a neutron that degrades into a proton, which becomes a semi-inclusionary proton within the negative atomic orbitals of an adjacent uranium atom. As a result of the atomic orbital distortion created by the semi-inclusionary proton, abnormal amounts of negative plasma breaches the area around the Neuproz Cluster, which causes expansion Neuproz expansion, dissonance, and subsequent decay.

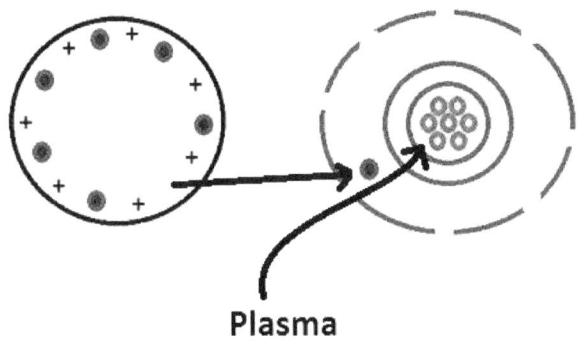

Plasma

Figure 22: The Slip

As you can see in this figure, a semi-inclusionary proton causes more plasma to inundate the atomic nucleus. All of which, is more logical: First, neutrons decay into protons really quickly; Second, atomic energy is ONLY useful in that it produces thermal energetic quanta to cause the expansion of water…to turn turbines. Therefore, in as much as old-theory purports that "neutron collisions" result in atomic fission, the temperature dependence of exponential fission means the METHOD of fission is based upon thermal energetic quanta disrupting the electronics within adjacent uranium atomic nuclei. (I don't know how they missed this logic, but you know…science is weird.) All of which, brings us back to variable electronic meshing and vibration.

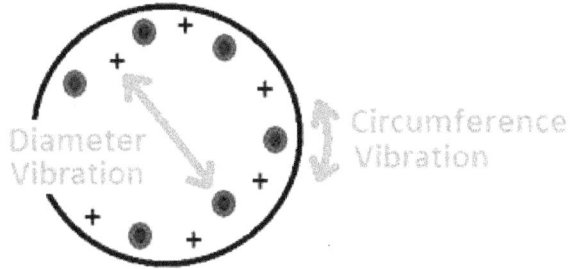

Figure 23: Vibration

As you can see in this figure, there are two types of vibration: Diameter and Circumference. Now, under normal circumstances, this is the hierarchy of vibrational importance:

1. Electronic meshing, diameter & circumference, dictates negative pillar vibration about the center of the Neuproz Cluster.
2. Negative pillar vibration dictates atomic orbital placement and wobble.
3. Atomic orbital placement and wobble dictates element vibration.

It's pretty straight forward and logical...if you ask me. But, under **extreme** circumstances, i.e. semi-inclusionary proton messing up atomic orbital placement, the hierarchy of vibrational importance inverts.

1. Atomic orbital wobble causes a distortion in negative pillar placement and vibration.
2. Negative pillar vibration results in different/weaker electronic meshing, i.e. diameter and circumference vibration.
3. Different vibrational patterns cause atomic particle extrusion, i.e. fission, based upon the AMOUNT of negative plasma present about the Neuproz Cluster.

As you can tell, things are complicated. But hopefully, they're more logical now. In any event, all of this leads to something very interesting: Macro-vibration.

In as much as extreme distortion of atomic orbitals with negative thermal energetic quanta and/or protons results in the enhancement of the isotopic degradation pathway, the long term vibrational patterns can also result in the enhancement of the isotopic degradation pathway. For example, look down.

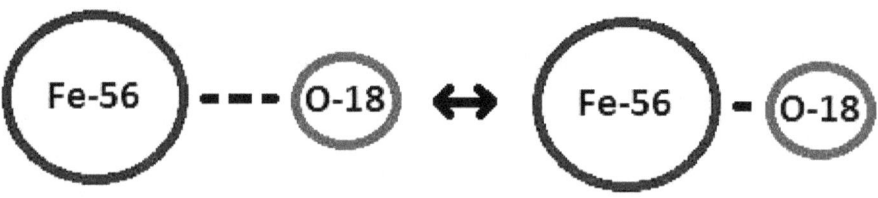

Figure 24: Shaky-Shaky

As you can see here, Fe-56 is more massive than O-18, which means the oxygen-18 does more of the MOVING. And as a result of the oxygen-18 moving more, the atomic orbitals wobble more. All of which, is the same reason why extreme negative thermal energetic quanta environments enhance the isotopic degradation pathway: More negative thermal energetic quanta reach the Neuproz Cluster. Or in simpler terms, do you notice anything different about this next figure?

^{26}Fe	**54** (5.8)	56 (91.72)	57 (2.1)	58 (0.28)		
^{25}Mn		55 (100.0)				
^{24}Cr	**50** (4.345)	52 (83.789)	53 (9.501)	54 (2.365)		
^{23}V	50 (0.250)	51 (99.750)				
^{22}Ti	**46** (8.0)	47 (7.3)	48 (73.8)	49 (5.5)	50 (5.4)	
^{21}Sc	45 (100.0)					
^{20}Ca	**40** (96.941)	42 (0.647)	43 (0.135)	44 (2.086)	46 (0.004)	48 (0.187)
^{19}K	39 (93.2581)	40 (0.0117)	41 (6.7302)			
^{18}Ar	**36** (0.337)	38 (0.063)	40 (99.6)			
^{17}Cl	35 (75.77)	37 (24.23)				
^{16}S	**32** (95.02)	33 (0.75)	34 (4.21)	36 (0.02)		
^{15}P	31 (100.0)					
^{14}Si	**28** (92.23)	29 (4.67)	30 (3.10)			
^{13}Al	27 (100.0)					
^{12}Mg	**24** (78.99)	25 (10.0)	26 (11.01)			
^{11}Na	23 (100.0)					
^{10}Ne	**20** (90.48)	21 (0.27)	22 (9.25)			
^{9}F	19 (100.0)					
^{8}O	**16** (99.762)	17 (0.038)	18 (0.2)			
^{7}N	14 (99.634)	15 (0.366)				
^{6}C	**12** (98.90)	13 (1.10)	14 (0.0)			
^{5}B	10 (19.2)	11 (80.2)				
^{4}Be	**9** (100.0)					
^{3}Li	6 (7.5)	7 (92.5)				

Figure 25: Isotopic Abundance of Lighter Elements (From *Incorporeal.*)

Now, I realize that Earth was purported to be made in all its glory in seven days. Butt, the data suggests, based upon the isotopic degradation pathway and the isotopes therein, that the Earth has been around long enough to evolve into a NON-radioactive environment. Or in other terms, the shaky-shaky of lighter elements has made them squirt out most of their extra neutrons. Or in carbon-dioxide terms, millions of years of carbon-dioxide vibration has rendered oxygen/carbon non-radioactive...thank God!

In conclusion, the structure of the elements in this branch of the universe is such that the negative electrons protect the positive charges within the atomic nucleus, e.g. Neuproz Cluster. In the event that vibration and/or extreme environments interfere with the atomic orbitals' ability to protect the Neuproz Cluster, enhance isotopic degradation will occur. As for the method of isotopic degradation, it is dependent on: Neuproz Groupings within the Neuproz Cluster; the age of each particle within the Neuproz Cluster; the diameter and circumference vibration of each particle; and, the type of negative energy innervating the Neuproz Cluster. And finally, based upon the variance of radiological decay of elements centered around negative thermal energetic quanta, the old theory of fission is wrong. Quite simply, neutrons degrade into protons, protons become semi-inclusionary, and negative thermal energetic quanta scoot past the atomic orbitals to cause the subsequent fission-ary events. (It's kind of weird how we can design and utilize such technology without truly understanding it?)

Chapter 11: Magnetism

If you've thought about, magnetism is weird. First and foremost, iron is weird because it has several oxidation states and also conducts electricity. And what is even weirder is: when you apply a current to or about iron, it becomes magnetic. In any event, after thinking about this, here are my thoughts on this unique matter.

In as much as the Neuproz Groupings are the "order" of the atomic nucleus and impart "order" to the atomic orbitals, there is a limitation to iron crystal formation because of thermal energetic quanta. But fortunately, as a result of this lack of crystal order, as well as variable oxidation states, a "free electron cloud" is permissive within iron alloys. All of which, is the method by which iron partitions into two distinct poles. Actually, let me try and explain a little bit better.

As a result of the Isotopic Degradation Pathway, Nickel-(58) probably degrades into Iron-(56) via the release of two protons, which means the asymmetry of Iron-(56) is the result of the inclusion of two protons on one side of the hemisphere as a result of the degradation of [5N] into a [3N]. All of which, can be described by my weird notation: 2(N)3[3N]2(N)2[5N][7N]. In any event, as a result of the lopsidedness created by the inclusion of two neutrons near the equator of iron on one side of the hemisphere and two neutrons near the axis of iron on the other side of the hemisphere, which is supported by the axis degradation of two neutrons from iron-(56) to form iron-(54), iron-(56) has some wobbly atomic orbitals. Or in simpler terms, wobbly atomic orbitals create magnetism and variable oxidation states, which produces "free electron cloud" in iron. And as a result of the "free electron cloud", there is a method to the force required to rotate iron atoms.

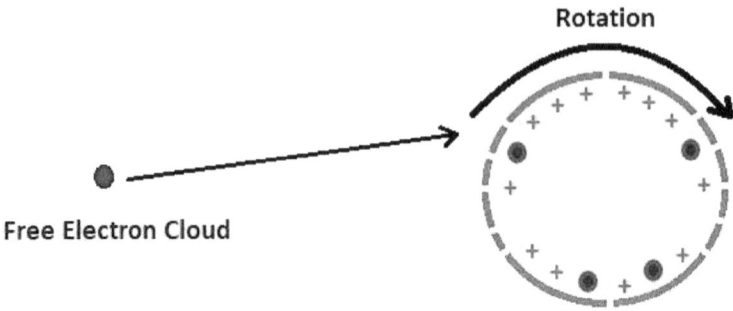

Figure 26: NON-free Rotation

As you can surmise from this figure, hopefully, is that due to the unsymmetrical nature of iron's atomic orbitals, certain regions of the sphere have more POSITIVE charge, which results in electrons smashing into these regions and rotating the atom. As for the AMOUNT of rotation, that depends on the electronic interaction with the adjacent atoms. All of which, brings us to the most important iron quandary: Why is iron bipolar? (Also, why didn't God create drugs to help iron with its bipolar disorder?) But first, a quick review.

1. Iron's Neuproz Cluster is asymmetric, which means iron's atomic orbitals are asymmetric.
2. As a result of this asymmetry, iron can exist in multiple oxidation states, which creates a "free electron cloud".
3. And finally, electrons pelting the asymmetric iron atom, results in rotation of the atoms to align into a bipolar arrangement based upon atomic orbital interactions.

Now, you HAVE to know by this point that I'm going to say that iron's bipolar arrangement is the result of energy conservation. Butt, how does the lattice structure of iron play a factor in magnetic strength? Also, how does the degradation of similar magnetic quanta facilitate the stabilization of NON-lattice iron?

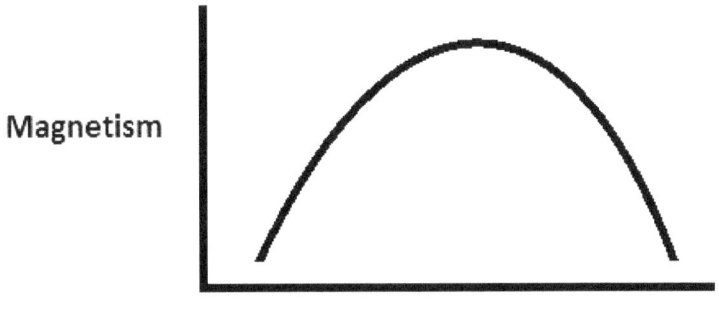

crystal : rotational atoms

Figure 27: The G-Spot, i.e. Gravity-Spot☺

As you can see in this figure, the strength of magnetism is based upon the ratio of crystal iron atoms to rotational iron atoms. Or in scientific terms, based upon the rate of cooling, doping agents, and external magnetic environment, some iron atoms become part of a crystal/lattice structure and some iron atoms become rotational active iron atoms. As for the method of bipolar energy conservation, it is quite simple...after understanding this complex figure.

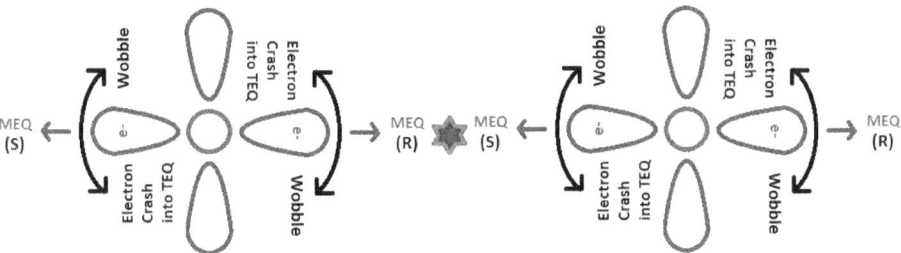

Figure 28: Rotational Atoms

The rotational alignment of iron atoms is the result of EXCESS negative plasma, which is notated by the blue and red star. And quite simply, the EXCESS blue and red stars protect the atomic nucleus from negative plasma attacking the atomic nucleus via the atomic orbital wobble zones.

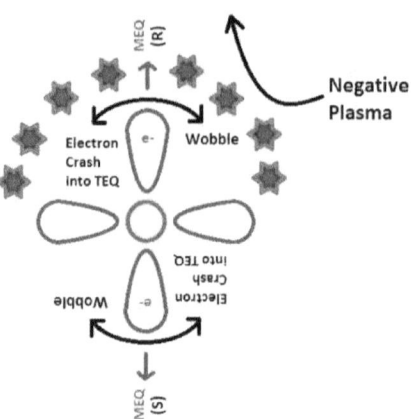

Figure 29: Plasma Cloud

As you can see in this figure, EXCESS blue and red stars, which are the result of opposite magnetic energetic quanta colliding and degrading, has formed a Plasma Cloud outside the atomic orbital wobble regions. And as a result of Plasma Could protecting negativity from attacking the atomic nucleus via the wobble regions, energy conservation occurs. Granted, this energy conservation is annoyingly small, but it is energy conservation none-the-less. Or in other terms, magnetism is simply PUMPING negativity away from the atomic nucleus and into Plasma Clouds, which protects the atoms on a micro scale. On a macro scale, magnetism creates a magnetosphere around the magnet. All of which is really interesting because the micro scale negative Plasma Cloud helps direct magnetism out of the magnet to create the macro Plasma Cloud, which is the one we detect. Or in simpler terms, the negative micro Plasma Cloud refracts negative magnetic energetic quanta out of the magnet to create the negative macro Plasma Cloud, e.g. magnetosphere.

On the off chance I wasn't completely clear on how lattice structure and rotational atoms relates to magnetism, let me introduce one more data point: glass-esk metals. When molten metal is cooled rapidly, the metal atoms don't have time to arrange into repetitive subunits, i.e. lattice structure. And as a result of this, there is NO 'hand of God' to stabilize rotational atoms to a bipolar arrangement.

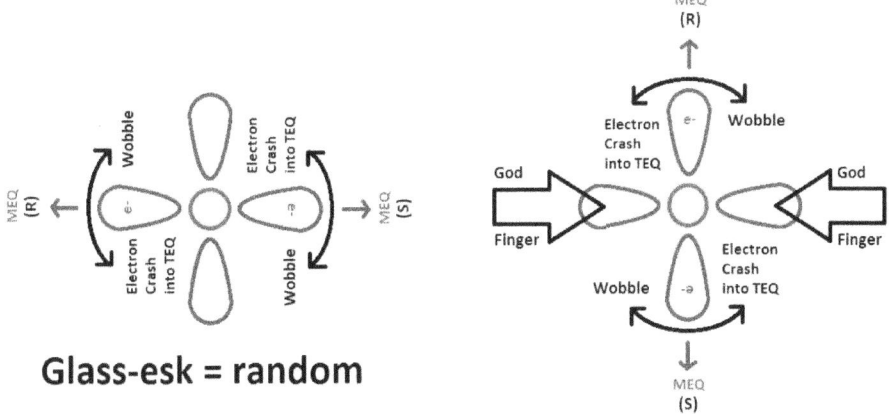

Figure 30: Hand of God

As you can see in this figure, God-finger-arrows, i.e. the order imparted by the atomic orbitals of an ordered lattice structure, holds the rotational iron atoms to a semi-bipolar arrangement. (I'll explain the 'semi-bipolar arrangement' concept in a moment.) Conversely, since there is NO ordered lattice structure within glass-esk iron, the lowest energy setting for the iron atoms is in reference to the external magnetic field, which does not create a strong bipolar arrangement. Or in simpler terms, glass-esk atoms exist randomly in 360°…based upon the external magnetic environment. Or in the simplest terms, lattice structures impart some level of order to rotational iron atoms.

Even though lattice structures limit the angle of rotation to create a bipolar magnet, the lattice structures themselves are not perfectly bipolar, which limits the strength of the magnet, i.e. semi-bipolar arrangement. Or in simpler terms, the three-dimensional lattice structure about the rotational atoms determines the atomic orbital wobble, which determines when, where, and with how much momentum the electrons crash into the thermal energetic quanta. (For a review of this concept, please see

Plinth:B.U.T.T. chapter 5.)   All of which, brings up a very interesting question: Is there a cross-lattice structural factor? Or in other words, does the association of lattice structures cause some iron atoms to release stronger magnetic energetic quanta?

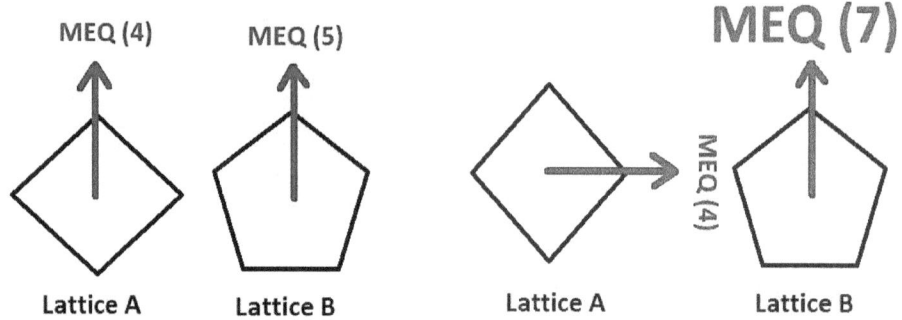

Figure 31: Lattice-Lattice interactions

As you can see in this figure, when Lattice A and Lattice B are parallel, their respective MEQs are 4 & 5. But when Lattice A is perpendicular to Lattice B, Lattice B releases a stronger MEQ (7).   All of which gets really complicated, really fast, since the world is a three-dimensional place with tons of different lattice structures.

In conclusion, magnetism is the cumulative result of all the factors that allow for the proximal association and focusing of non-lattice bound rotationally-free iron atoms. Methodology, the oxidation state of iron results in a free electron cloud, which allows electrons to bombard asymmetrically positive iron atoms until they rotate to align their opposite magnetic energetic quanta.   Once their opposite magnetic energetic quanta align, the degradation of the opposite magnetic energetic quanta results in the negative plasma cloud, which inhibits the flow of negative plasma towards the wobbling atomic orbitals and directs negative magnetic energetic quanta out of the magnet to create a macro plasma cloud. All of which, is a coarse function of crystal-to-rotational atoms in the absence of: nano-technology, expensive doping agents, and/or magnetically facilitating crystal-lattice alignment during the cooling process.

Chapter 12: Sustained Movement

We've seen it a hundred times: a compass aligning to the Earth's magnetosphere. Butt, since movement requires force and force requires energy, the Earth's magnetosphere is comprised of energetic quanta. (I never thought the justification behind magnetic energetic quanta would be that simple to explain...butt, I guess it was.) In any event, I've been wondering: Could we build tiny motors powered by Earth's magnetosphere?

If you or one of your ancestors was a successful thinker, you might have one of those expensive watches that are charged based upon your arm's movement. As for the pendulum-esk nature of this expensive time-keep, it's interesting. But what would be more interesting is: a magnetic-shell-matrix that uses Earth's magnet flux to facilitate the movement of a magnet to produce energy. Or in simpler terms, a pendulum that uses the Earth's magnetosphere, i.e. gravity, to sustain perpetual movement.

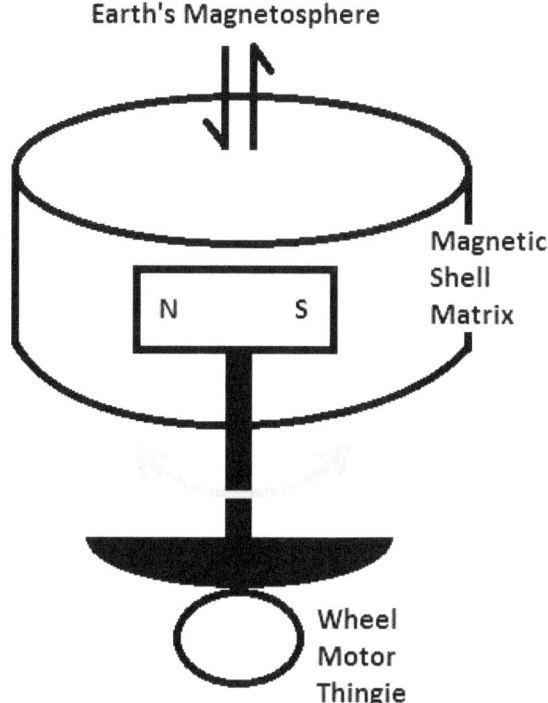

Figure 32: Sustained Movement

As you can surmise from this figure, hopefully, is that the 'magnetic shell matrix' creates a magnetic flux, which forces the bar magnet to align. And as a result of this continual force, sustained movement is accomplished, which will make this device look like a never-ending pirate-boat ride at Cedar Point...or whatever amusement park you associate with pirate-boat rides. In any event, part of this sustained movement can be converted into energy if a wheel/motor is in contact with the swinging pendulum. In theory, the strength of the magnet as well as the 'magnetic shell matrix' will counter act the resistance caused by: Gravity, friction, and spinning the wheel attached to a motor.

The easiest way to imagine it is: the bar-magnet is the magnesium at the center of the polyporphyrin of chlorophyll, which is the 'magnetic shell matrix'. The energy of Earth's magnetosphere causes the magnesium to wobble until it releases a high energy electron, i.e. sustained pendulum movement.

In conclusion, science has the propensity to make the world a better place. All we have to do is get the theories a little more correct, give the next generation a little nudge, and hope for the best. All of which, might result in a magnetic pendulum the size of the Eiffel Tower...continually producing energy and attracting tourists from around the world.

Chapter 13: Flint Vs. Lightning

This next quandary goes out to the people of Flint Michigan and everybody without fresh water...since I have no money to give. For those of you who don't know this, flint is a type of iron that sparks when it's lattice structure is cracked, which provided early humans with fire. Unfortunately today, flint is still used by humans to inhale the smoke & the pain-killer known as nicotine...even when they could switch to Blu and live a longer, happier, and better life. In any event, flint is very interesting.

Now at first glance, pun intended, it was thought that electrons were jumping outward from the cracked flint, like lighting. But, in truth, flint reacts with oxygen about a BALANCED equation. All of which means, the light released by this reaction is the result of negative plasma fluxing towards the flint...probably because rearrangement of iron's atomic orbitals, which result in more 'displayed' positivity. Or in simpler terms, look down.

Cloud (e-)  Earth (+)

Figure 33: Lightning-Bolt

As you can see in this figure, the negative electrons in the clouds are moving towards Earth's relatively positive crust. Therefore, the

MOVEMENT of **electrons** cause the distortion of atomic orbitals such that electrons crash into plasma, degrade, and release photons of light.

Flint (+) **Plasma (-)**

Figure 34: Flint-Bolt

Now flint on the other hand, causes the MOVEMENT of **plasma** towards the flint's relatively positive crust. Or in Quanta Dynamic terms, the momentary structurally facture of the elements in the flint results in atomic orbital re-orbitalization, which increases the 'observable' positivity within some of the flint atoms. And as such, negative plasma moves TOWARDS the atoms of flint, which results in the distortion of atomic orbitals and the release of light by the degradation of negative electrons.

As for the reason why both bolts of energy are organized oppositely, i.e. fat-end towards the positive-flint and skinny-end towards the positive-Earth, it is quite simple. In the case of the lightning bolt, the closer the electron bolt gets to the positive-Earth, the more diffuse the electrons become, which makes a less concentrated electron flow to distort atomic orbitals and cause the release of light. In the case of the flint-bolt, the further away from the positive-flint, the less negative plasma is pulled towards the positive-flint, which means the skinny-end of the flint-bolt is facing away from the flint.

In conclusion, the theory is exactly the same between lightning-bolts and flint-bolts, i.e. electrons crashing into plasma to release photons of light. But, the method differs in each case: lightning moves electrons to distort atomic orbitals, which causes the collision of electrons into plasma; flint-bolts are the result of plasma surges towards relatively positive flint, which causes the collision of electrons into plasma. In any event, best of luck finding clean water.

Chapter 14: Hail-O

After thinking about hail for several days, the current theory of hail formation just doesn't make any sense. I mean, if you believe that matter can neither be created/destroyed and that Earth is a closed system, then sure...hail is formed because of wind. But, when you factor in the Isotopic Degradation Pathway, a much more interesting theory of hail formation appears.

To begin with, let's review my logic thus far. First, we're in a negative branch of the universe. Second, positive matter has mass because of the presence of excess negative energy. Next, photons, magnetic energetic quanta, thermal energetic quanta, and plasma are **negative**...in this negative branch of the universe. And finally, because we're in a negative branch of the universe, negative energy diffuses outwards.

Now, with this review in mind, if a positive proton degraded and released MOST of its positive components, which will react with and destroy negative thermal energetic quanta, then the region of this 'reaction' will be cold because there will be LESS negative thermal energetic quanta.

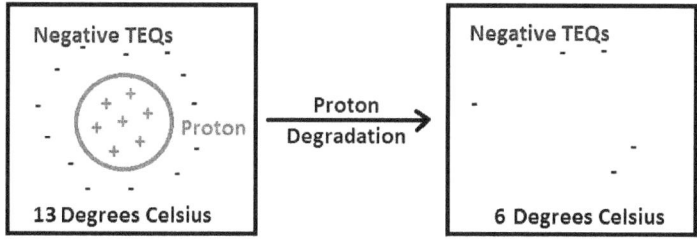

Figure 35: Proton Degradation

As you can hopefully surmise from the figure on the previous page, the 7-positive charges released by the degradation of the proton react with 7-negative thermal energetic quanta to produce 'thermally inactive plasma'. And as a result of this, the formal temperature, notated by the number of negative thermal energetic quanta drops from 13 to 6 degrees Celsius. All of which, causes the diffusion of negative thermal energetic quanta away from water, i.e. freezing, to form hail.

Now, even though this is insanely straight-forward with regards to Quanta Dynamics, I still have to explain 'thermally inactive plasma'. Quite simply, a negative thermal energetic quantum has MORE negative components than a plasma quantum. And as a result of the attraction between positive and negative being inversely related to distance, itsy-bitsy-tiny-weany plasma quantum has such a small amount of negativity, it diffuses instead of being gravitationally active. Or in simpler terms, the charge ratio between a thermal energetic quantum and a plasma quantum dictates that the thermal energetic quantum is gravitational and the plasma quantum is diffusional.

In conclusion, due the fact that protons are continually degrading to release negative magnetic energetic quanta, the magnetosphere of the atmosphere is funkier, Earth is continually diffusing negative plasma, and clouds release copious amounts of negative electrons back to Earth, proton degradation in the clouds destroys negative thermal energetic quanta, which freezes water to form hail. As per the size of the hail, that would probably be determined by the symmetry of the proton degradation into "more perfect" positive ANTI-thermal energetic quanta. Or in simpler terms, the more perfect the destruction of the proton into ANTI-thermal energetic quanta, the more negative thermal energetic quanta will be destroyed, which results in a colder locale to freeze more water, i.e. bigger hail.

Chapter 15: Computing

With the knowledge that atomic orbital wobbles can cause the surprising movement of plasma, as is the case with flint, then it shouldn't be a surprise that one day, in the not so distant future, someone will INVENT quantum computing atoms, which can be modulated to store a one or a zero, based upon the release of a photon. Luckily, if you can't imagine that, then I've got just the chapter for you. But first, let's do a quick review with regards to chlorophyll.

Now, technically, chlorophyll would NOT be classified as a quantum computing entity because it takes thousands of amino acids, a polyporphyrin ring, and magnesium to convert **multiple** photons into a high energy electron, but that shouldn't discourage us. (This concept is better explained in *Quanta Dynamics*.) Or in simpler terms, nature has evolved based upon energy efficiency, but humans lack that imagination. Or in the simplest terms, humans can imagine complexity NOT BASED upon energy efficiency. (Take of that what you will.) In any event, let's use chlorophyll as the basic idea for a quantum computing entity.

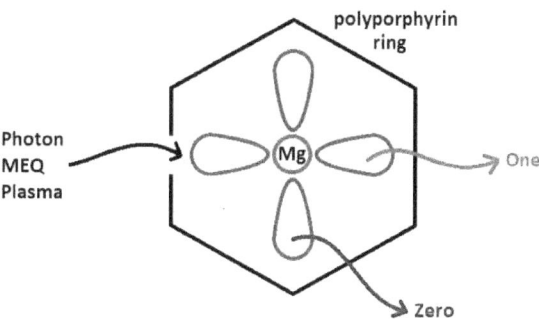

Figure 36: First Idea

In theory, quantum computing would involve the programing of an atom's atomic orbitals such that the stimulation with a photon, MEQ, or plasma will illicit the directional response of 'one' or 'zero'. Unfortunately, since atomic orbitals are a function of the environment, this will never lead to SINGLE atom quantum computing. But, it will provide smaller computing devices.

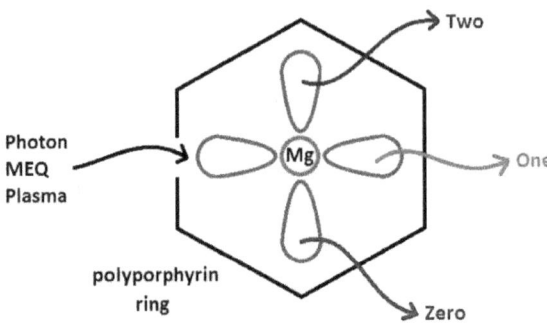

Figure 37: Second Idea

As you can see in this next scenario, directional release based upon atomic orbitals will result in the tertiary system, which will confuse the fuck out of binary coders and software developers...or so my brother told me nine years ago. But, don't get too excited yet about this possibility just yet. Even though the inclusion of a NON-binary system will counter act the space required for the polyporphyrin ring and may reach the level of 'quantum computing', single photon detection is a bitch. And DIRECTIONAL single photon detection is insane. AND...directional single photon detection through a non-refracting computing matrix will cause your head to explode! So, in as much as I'd love to continue on this road of cognition, I think there are easier methods to reducing the size of data storage, i.e. ones and zeros.

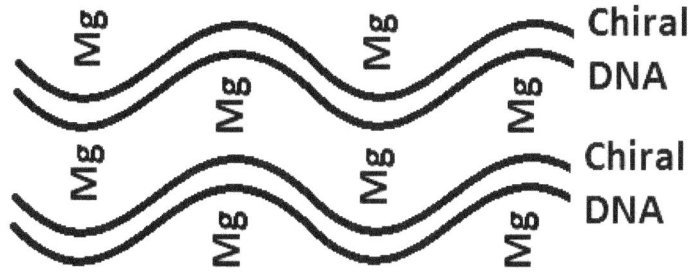

Figure 38: Third Idea

Now, in this third idea, there is an accumulation of concepts. First, different Chiral DNA segments will innervate and tune magnesium to different wavelengths, which will facilitate the distinction between layers, i.e. one layer responds to blue light and another to red light. Next, until laser focusing technology and single photon detection improves, multiple magnesium subunits can represent a "single" data storage unit. And finally, brewing chiral DNA is pretty simple.

In conclusion, my idea of for quantum computing doesn't really reach the level of quantum computing. But on the bright side, quantum computing might be a unicorn because the storage of data based upon atomic orbital position requires ADDITIONAL atoms to 3-dimenstionally innervate the quantum computing unit…unless NASA has made some ultra-pure crystals in space and are keeping their computing power a secret? In any event, there is still a lot to learn with regards to HOW each atomic orbital degrades photons or magnetic energetic quanta into plasma. And finally, nobody is doing research on HOW much plasma is required to cause a shift in unique atomic orbitals…either store data or retrieve data. But, we can dream…unless they take that away from us too…ahhhh, the horror!

Chapter 16: How About Them Apples?

So I've been doing some thinking: I MIGHT be able to resolve the difference between scientists and climate-deniers…with little insult to both. But, before I do all that and a bag of chips, let me be completely clear about one thing: Earth is going to kill us. The only true question is: At what RATE?

First and foremost, there is blame on both sides of the isle…scientists and climate-deniers. For example, the current theory of climate change is that heat is trapped by excess amount of carbon dioxide in the air, which is akin to the theory that hot air balloons NEVER release thermal energy and return to Earth. Actually, let me try and explain that a little better. Scientists believe that Earth is mathematical **closed** system and that more carbon dioxide will trap more heat. Now, in as much as I LOVE science and I would marry it if I could find it, carbon dioxide is ONLY a secondary function to climate change. Or in simpler terms, carbon dioxide ONLY matters as it determines the amount of carbonic acid in the ocean, which provides more available POSITIVE protons. Or in the simplest terms, since heat is NEGATIVE thermal energetic quanta, more POSITIVE protons in the ocean decreases the diffusion of NEGATIVE thermal energy quanta. Therefore, year after year, LESS thermal energy diffuses from the ocean, which causes the polar ice caps to melt.

So let me recap thus far: Excess carbon dioxide in the air has a MINISCULE effect on trapping itsy-bitsy-tiny-weeny negative thermal energetic quanta, mostly because carbon dioxide has a very small 'available' positive charge to attract negative thermal energetic quanta. In fact, if you think about it, the relative partial negativity about the two oxygen atoms probably REPELLS negative thermal energetic quanta. (FYI, that was a very moot point.) In any event, carbon dioxide dissolves into the oceans to form carbonic acid, which creates more NAKED positive protons...the TRUE culprit in ocean temperatures rising. And as a result of ocean temperatures rising, more water diffuses into the atmosphere, i.e. water has two mostly naked positive protons, which attracts more negative thermal energetic quanta.

Now, to the climate-deniers. Of course Earth goes through cycles of warm and cold periods...That's a no brainer. But, what climate-deniers are forgetting is: Rocks. (I know this sounds stupid, but just try and keep reading.) Another effect of carbonic acid within salt water is to facilitate the precipitation of sodium chloride, which acts as a buffer to changes in carbon dioxide levels. Or in simpler terms, when the Earth was younger, more volcanos released their salty goodness into the oceans, which forced carbon dioxide out of the water. Or in cyclic terms, carbonic acid equals warmer period and underwater volcanic eruptions result in the degassing of the ocean, which results in a cooler period. Unfortunately, Earth is getting OLD, which means it has ISLANDS. Or in weirder terms, ISLANDS release eruptions into the air instead of the water, which means islands DECREASE the amount of ocean degassing. And on top of all that complexity, there is the variance of elements being released by Earth's older core.

In conclusion, let me speak to both groups. First, the climate-deniers: Of course the Earth goes through cycles, but based upon age of the planet and the formation of islands, variable release of elements into the atmosphere in comparison to the ocean has a DIRECT relation as to part of the CYCLE where Earth heals itself. Or in simpler terms, young planets can heal themselves quicker because they vent their farts into the ocean. Conversely, old planets take MUCH longer to heal because they vent their farts into the atmosphere, which totally fucks up photosynthesis. (BTW, eventual oxidation of carbon to produce carbon dioxide is inevitable, but humans have modified that RATE, which means we are modulating the length of Earth's 'cooler' period with lower ocean levels.) Now, for the scientists: Protons are positive and thermal energetic quanta are negative. Therefore, more available POSITIVE protons in the ocean will decrease the rate of NEGATIVE diffusion as well as modulating the precipitation of the salt-buffer system. And finally, to everyone that has a vested interest in humanity being around for a couple more decades, the cold-hot cycles on Earth are based upon WHERE volcanic eruptions take place. Or in the simplest terms, OLD planets don't heal the same as YOUNG planets...a duh!

Chapter 17: XX^A

In the event that humanity develops the ability to compromise, which is highly unlikely, we're still probably going to argue about evolution. But, before you go thinking I'm on the side of science, let me be honest with you. The only thing standing between Mary producing a boy-offspring without having sex is: A fractured 'X' chromosome. (You'll understand that a little better...later in the chapter.) In any event, hopefully I can bridge the divide between the religious and the scientific-religious...or just piss them both off☺

In the beginning, there was one cell, which I already talked about in *Incorporeal*. Then, there were two cells. Unfortunately, as a result of there being two cells and the world being a variable 3-dimensional entity, each cell got exposed to a different environment. And as a result of the different environment, and all the molecules therein, different proteins were enhanced and/or diminished, which resulted in the variable expression of the genome. Or in simpler terms, a cell exposed to DIFFERENT molecules will express different genes. And so began the miracle of life: the association of cells based upon a competitive advantage created by some level of synergy, i.e. one cell might eat disaccharides and poop out mono-saccharides, which the adjacent cell might like. In any event, to make a very long story much shorter, variable expression based upon environment and the need for some level of synergy resulted in a multi-cellular organism, which is where things get interesting.

Let's hypothetically say that the first multi-cellular organism 'XX^A' was an asexual entity that laid eggs, i.e. NO need for egg fertilization.

Unfortunately, based upon the environment and very long periods of time, more and more eggs were born with fractured genomes, i.e. 'XY'. Or in simpler terms, the first male entities were accidents and did NOT play a part in reproduction, but they provided some BENEFICIAL evolutionary characteristic to the asexual 'XX^A' parent. For example, let's say that the mutant 'XY' entity was severely thick headed, had more muscle, and an insane desire to protect & squirt out sperm. All of which was completely useless until a mutant 'XX' entity REQUIRED the sperm stimuli to facilitate growth of new entity. Or in simpler terms, the 'male' entities could have been a very common sporadic event for millions of years because they provided some evolutionary benefit to the asexual female host 'XX^A', but the TRUE two-part system of 'male' and 'female' reproduction ONLY occurred when a mutant 'XX' occurred, which required additional stimuli from an 'XY' partner to produce offspring. But, what is the relationship to between the mutant 'XX' and 'XY' that was beneficial to variable environmental stresses?

In order to answer this question, let's think about homicidal spiders, sperm, and chemotaxis. Now for those of you who don't know this, there are spider-woman that after being fertilized by spider-man, the spider-women eat spider-man. Maybe it is because the spider-men have more protein in their bodies or simply because they're bad lovers. All of which leads us to the next idea: sperm. Hopefully, everybody has seen the Homer-sperm episode of the Simpsons...mostly because it will make this next part much easier to imagine. In any event, when the Homer-sperm are trapped in a man's testes, they're fed plenty of carrots and NO beer, which makes them extremely docile. But after being released into Marg's vaginal canal, the Homer-sperm swim around randomly until they get a whiff of Marg's beer-egg, which brings us to chemotaxis. (That totally puts a spin on yeast infections...ewe!)

The basis of the word 'chemo-taxis' is the movement of an object towards a chemical, which is somewhat confusing since nobody associates chemicals with energy. It would have been a whole lot easier to understand if they just called it energy-taxis...but that is just me. In any event, sperm move towards chemical energy, which is being released by the beer-egg. All of which, brings to mind something my brother told me years ago, i.e. more than nine years ago.

A long time ago, before politics destroyed our relationship, my brother told me that female sperm mature SLOWER than male sperm. Actually, my brother said: if you want to have a girl, don't ejaculate for a couple days before trying to conceive. In any event, the reason why this is interesting, and somewhat relevant, is because sperm CHEW their way into beer-eggs. Or in simpler terms, you remember how sperm move towards chemical energy? Well, sperm don't have mouths to chew up big molecules. But, sperm does secrete enzymes to break down chemical energy so that it can be absorbed through their membranes. And ironically, these enzymes play a part in the sperm CHEWING their way into the beer-egg.

In any event, here is my theory of mutual beneficial evolution: Female beer-eggs have a protein deficiency that is relieved by eating the outside of the idiot-sperm, which ironically ate its way into the beer-egg. (Eating your way to death...sounds like a bad movie title.) Or in other words, female beer-egg digests the protein in the sperm's body, which relinquishes the protein deficiency to begin mitosis. All of which brings us to the REASON why this two-part system evolved. Or in simplest terms, more food equals more sperm, which equals more off-spring.

As for the genetic mis-matching between egg and sperm? Well, that is probably a relic of the plasmid wars between bacteria, which somehow got reactivated based upon better survival characteristics.

In conclusion, by working backwards from known genetic fractures, i.e. Down Syndrome, with the knowledge of twin events, it is plausible to postulate that asexual 'XX^{A}' multi-cellular entities produced 'XY' entities, which had NO sexual purpose, but provided some evolutionary benefit, i.e.

protection. All of which, could have been going on for millions of years until a mutant 'XX' occurred that required sperm-protein to relieve a protein deficiency to produce feasible offspring in an energetically conducive environment.   And from there, lots of mutant bunny generations resulted in different evolutionary tracts based upon external stimuli and genome swapping.   Or in simpler terms, 'XX^A' accidently produced 'XY', but 'XY' was TRULY irrelevant until 'XX' evolved to require the 'XY' in the sex process...I think. All of which, is just as hard to believe as Mary's 'XX' chromosome fracturing to produce Jesus's 'XY' genetic makeup...without having sex. (BTW, based upon the way men "used" to treat women, it would not have been surprising if Joseph punched Mary in the gut for not putting out, which might have facilitated the genomic fracture...I'm trying to keep it real people!)

Chapter 18: Contingent Beliefs

The Golden Rule **before** the FIRST Golden Rule is: You gotta do what you gotta do to keep your head from hurting. All of which, results in contingent beliefs. In the event your beliefs haven't allowed you to prosper in whatever realm of imagination you like to dabble in, let's take a moment and discuss what it is you imagine. Do you imagine people are inherently good? Do you imagine that going through the pain of childrearing will make your husband believe that you love him? Do you believe you've been inherently shorted by everything? If so, you're not alone.

In my mental exploration, I have yet to come across any explanation for what people seek and how they go about seeking it. People are truly an amalgam of the things they've experienced and imagined. So how do you go about finding people that care about you as much as you want to care about them? Well, the old-school method was to pop-out a kid and add-time to create love. And the new-age belief is a highly variable function of personal psychiatric maladies, which make you wet at night. In any event, all of this confusing hoopla is the reason I began thinking about science. Quite simply, I couldn't find someone that 'loved' me, so I decided to 'love' science.

After realizing that science loved me back, I needed a reason past science. (Science typically uses abstinence as a vindictive method of control...or that is just a current evolutionary problem?) Therefore, I made the unholy combination of my 'opinion' with my scientific theories. Does this make my opinions better or more-right? Probably not. Is my belief less of a BELIEF if 'empathy' can't make the world a better place? Not in the least.

But, in the event that my first belief isn't correct or conducive to anything other than my imagination, I must call upon my contingent beliefs. Now, I just have to figure out what they are and how they will affect the world...if I still care about all that hoopla. Also, I would like to have sex again before I die...it was nice.